Editor
Lorin Klistoff, M.A.

Managing Editor
Karen Goldfluss, M.S. Ed.

Editor-in-Chief
Sharon Coan, M.S. Ed.

Cover Artist
Barb Lorseyedi

Art Manager
Kevin Barnes

Art Director
CJae Froshay

Imaging
Rosa C. See

Product Manager
Phil Garcia

Publisher
Mary D. Smith, M.S. Ed.

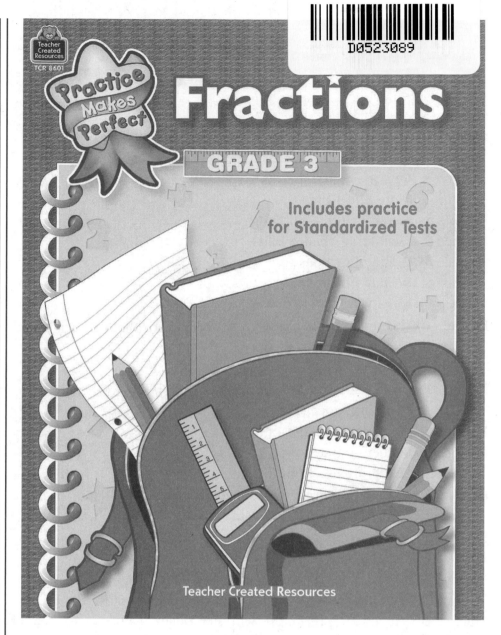

Fractions

GRADE 3

Includes practice for Standardized Tests

Practice Makes Perfect

TCR 8601

Teacher Created Resources

Author

Mary Rosenberg

Teacher Created Resources, Inc.
12621 Western Avenue
Garden Grove, CA 92841
www.teachercreated.com

ISBN: 978-0-7439-8601-4

©2004 *Teacher Created Resources, Inc.*
Reprinted, 2018
Made in U.S.A.

Table of Contents

Introduction 3
Practice 1: Identifying Parts 4
Practice 2: Identifying Equal Parts 5
Practice 3: Identifying Equal Parts 6
Practice 4: More Equal Parts 7
Practice 5: Identifying Shaded Parts 8
Practice 6: Identifying Shaded Parts 9
Practice 7: Writing Fractions 10
Practice 8: Writing Fractions 11
Practice 9: More Writing Fractions 12
Practice 10: Naming Proper Fractions as Part of a Whole 13
Practice 11: Naming Proper Fractions as Part of a Whole 14
Practice 12: Shading Parts 15
Practice 13: Writing Fractions Using Numbers and Words 16
Practice 14: Naming Proper Fractions as Part of a Set 17
Practice 15: Finding the Largest Fraction 18
Practice 16: Ordering Fractions 19
Practice 17: Ordering Fractions 20
Practice 18: Identifying Fractions Using a Number Line 21
Practice 19: Identifying Fractions Using a Number Line 22
Practice 20: Identifying Fractions Using a Number Line 23
Practice 21: Adding Fractions with Like Denominators 24
Practice 22: Adding Fractions with Like Denominators 25
Practice 23: Adding Fractions with Like Denominators 26
Practice 24: Subtracting Fractions with Like Denominators 27
Practice 25: Subtracting Fractions with Like Denominators 28
Practice 26: Subtracting Fractions with Like Denominators 29
Practice 27: Adding and Subtracting Fractions with Like Denominators 30
Practice 28: Comparing Fractions 31
Practice 29: Comparing Fractions 32
Practice 30: Comparing Fractions 33
Practice 31: Naming and Comparing Fractions 34
Practice 32: Finding Equal Sets 35
Practice 33: Finding Equal Sets 36
Practice 34: Finding Part of a Whole 37
Practice 35: Representing Money as Fractions 38
Practice 36: Representing Money as Fractions 39
Practice 37: Graphing Fractions 40
Test Practice Pages 41
Answer Sheet 46
Answer Key 47

The old adage "practice makes perfect" can really hold true for your child and his or her education. The more practice and exposure your child has with concepts being taught in school, the more success he or she is likely to find. For many parents, knowing how to help your children can be frustrating because the resources may not be readily available. As a parent it is also difficult to know where to focus your efforts so that the extra practice your child receives at home supports what he or she is learning in school.

This book has been designed to help parents and teachers reinforce basic skills with their children. *Practice Makes Perfect* reviews basic math skills for children in grade 3. The math focus is on fractions. While it would be impossible to include all concepts taught in grade 3 in this book, the following basic objectives are reinforced through practice exercises. These objectives support math standards established on a district, state, or national level. (Refer to the Table of Contents for the specific objectives of each practice page.)

- identifying equal parts
- identifying shaded parts
- writing fractions
- naming proper fractions as part of a whole
- shading parts
- writing fractions using numbers and words
- naming proper fractions as part of a set
- finding the largest fraction
- ordering fractions

- identifying fractions using a number line
- adding fractions with like denominators
- subtracting fractions with like denominators
- comparing fractions
- finding equal sets
- finding part of a whole
- representing money as fractions
- graphing fractions

There are 37 practice pages organized sequentially, so children can build their knowledge from more basic skills to higher-level math skills. (*Note:* Have children show all work where computation is necessary to solve a problem.) Following the practice pages are five test practices. These provide children with multiple-choice test items to help prepare them for standardized tests administered in schools. To correct the test pages and the practice pages in this book, use the answer key provided on pages 47 and 48.

How to Make the Most of This Book

Here are some useful ideas for optimizing the practice pages in this book:

- Set aside a specific place in your home to work on the practice pages. Keep it neat and tidy with materials on hand.

- Set up a certain time of day to work on the practice pages. This will establish consistency. An alternative is to look for times in your day or week that are less hectic and more conducive to practicing skills.

- Keep all practice sessions with your child positive and constructive. If the mood becomes tense or you and your child are frustrated, set the book aside and look for another time to practice with your child.

- Help with instructions, if necessary. If your child is having difficulty understanding what to do or how to get started, work through the first problem with him or her.

- Review the work your child has done. This serves as reinforcement and provides further practice.

- Allow your child to use whatever writing instruments he or she prefers. For example, colored pencils can add variety and pleasure to drill work.

- Pay attention to the areas in which your child has the most difficulty. Provide extra guidance and exercises in those areas. Allowing children to use drawings and manipulatives, such as coins, tiles, game markers, or flash cards, can help them grasp difficult concepts more easily.

- Look for ways to make real-life applications to the skills being reinforced.

Practice 3

Directions: Read each question. Then circle the correct shapes.

---

1. Which shapes are divided into **four** equal parts?

---

2. Which shapes are divided into **three** equal parts?

---

3. Which shapes are divided into **two** equal parts?

---

4. Which shapes are divided into **five** equal parts?

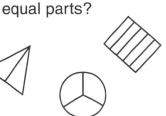
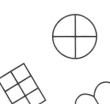

---

5. Which shapes are divided into **six** equal parts?

Practice 4

Directions: Divide each shape into equal parts.

1. 8 equal parts 	**2.** 4 equal parts 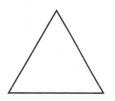	**3.** 3 equal parts 
4. 2 equal parts 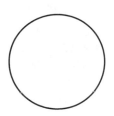	**5.** 5 equal parts 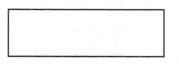	**6.** 2 equal parts 
7. 5 equal parts 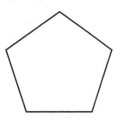	**8.** 4 equal parts 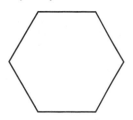	**9.** 3 equal parts 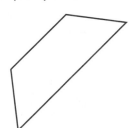
10. 3 equal parts 	**11.** 4 equal parts 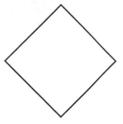	**12.** 5 equal parts 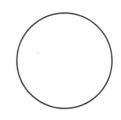
13. 2 equal parts 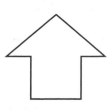	**14.** 6 equal parts 	**15.** 4 equal parts 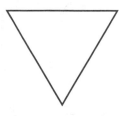

Practice 5

Directions: How many parts are shaded? Write the number.

1.

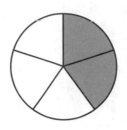

_____ parts

2.

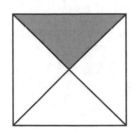

_____ part

3.

_____ part

4.

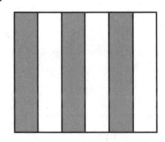

_____ parts

5.

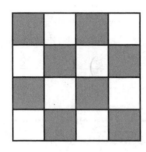

_____ parts

6.

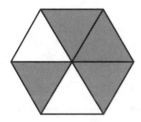

_____ parts

7.

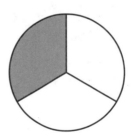

_____ part

8.

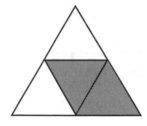

_____ parts

9.

_____ part

10.

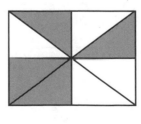

_____ parts

11.

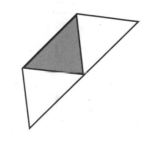

_____ part

12.

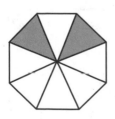

_____ parts

Practice 6

Directions: Write the number of shaded circles in the box.

1. □/6	2. □/7
3. □/4	4. □/6
5. □/10	6. □/9
7. □/2	8. □/3
9. □/9	10. □/8
11. □/8	12. □/9
13. □/5	14. □/8
15. □/3	16. □/4

Practice 9

A **fraction** is a number that names part of a whole thing. The number at the top is the numerator. It tells how many parts of the whole are present. The number at the bottom is the denominator. It tells how many parts there are in all.

Examples

 $\frac{1}{2}$ (There are two parts in the circle. One part is gray. Therefore, the fraction is $\frac{1}{2}$.)

 $\frac{3}{4}$ (There are four parts in the square. Three parts are gray. The fraction is $\frac{3}{4}$.)

Directions: Write a fraction for each picture.

1. 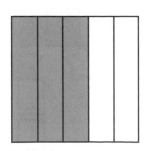 _____

2. 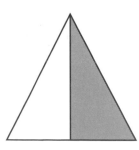 _____

3. 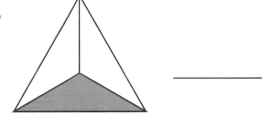 _____

4. 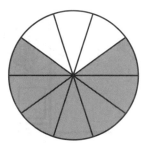 _____

5.  _____

6. 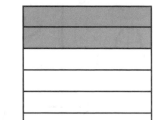 _____

7. 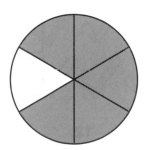 _____

8. 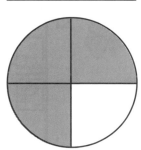 _____

Practice 10

Directions: Read each question. Then circle the correct answer.

1. What fraction of the circle is shaded?

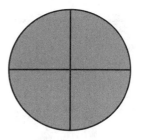

(A) $\frac{5}{2}$ (B) $\frac{4}{1}$ (C) $\frac{5}{4}$ (D) $\frac{4}{4}$

2. What fraction of the circle is shaded?

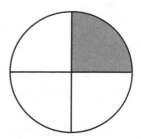

(A) $\frac{2}{4}$ (B) $\frac{1}{4}$ (C) $\frac{4}{1}$ (D) $\frac{2}{2}$

3. What fraction of the circle is shaded?

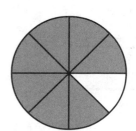

(A) $\frac{4}{8}$ (B) $\frac{7}{8}$ (C) $\frac{8}{7}$ (D) $\frac{8}{4}$

4. What fraction of the circle is shaded?

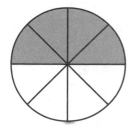

(A) $\frac{5}{4}$ (B) $\frac{4}{5}$ (C) $\frac{4}{8}$ (D) $\frac{8}{4}$

5. What fraction of the circle is shaded?

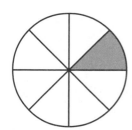

(A) $\frac{2}{4}$ (B) $\frac{4}{2}$ (C) $\frac{1}{8}$ (D) $\frac{8}{1}$

6. What fraction of the circle is shaded?

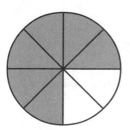

(A) $\frac{7}{4}$ (B) $\frac{6}{8}$ (C) $\frac{8}{6}$ (D) $\frac{4}{7}$

Practice 11

Directions: Read each question. Then circle the correct answer.

1. What fraction of the rectangle is shaded?

   (A) $\dfrac{3}{7}$ (B) $\dfrac{3}{4}$ (C) $\dfrac{4}{7}$ (D) $\dfrac{1}{7}$

5. What fraction of the rectangle is shaded?

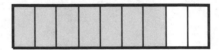

   (A) $\dfrac{1}{9}$ (B) $\dfrac{7}{9}$ (C) $\dfrac{2}{9}$ (D) $\dfrac{7}{2}$

2. What fraction of the rectangle is shaded?

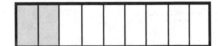

   (A) $\dfrac{2}{9}$ (B) $\dfrac{2}{7}$ (C) $\dfrac{7}{9}$ (D) $\dfrac{1}{9}$

6. What fraction of the rectangle is shaded?

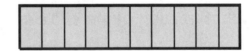

   (A) $\dfrac{1}{10}$ (B) $\dfrac{10}{10}$ (C) $\dfrac{0}{10}$ (D) $\dfrac{10}{0}$

3. What fraction of the rectangle is shaded?

   (A) $\dfrac{4}{9}$ (B) $\dfrac{1}{9}$ (C) $\dfrac{4}{5}$ (D) $\dfrac{5}{9}$

7. What fraction of the rectangle is shaded?

   (A) $\dfrac{5}{7}$ (B) $\dfrac{1}{7}$ (C) $\dfrac{2}{7}$ (D) $\dfrac{5}{2}$

4. What fraction of the rectangle is shaded?

   (A) $\dfrac{0}{6}$ (B) $\dfrac{6}{0}$ (C) $\dfrac{1}{6}$ (D) $\dfrac{6}{6}$

8. What fraction of the rectangle is shaded?

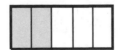

   (A) $\dfrac{3}{5}$ (B) $\dfrac{1}{5}$ (C) $\dfrac{2}{3}$ (D) $\dfrac{2}{5}$

Practice 12

Directions: Shade the parts to match the fraction.

1. $\frac{1}{7}$	**2.** $\frac{2}{5}$	**3.** $\frac{3}{8}$
4. $\frac{4}{9}$	**5.** $\frac{4}{7}$	**6.** $\frac{1}{3}$
7. $\frac{7}{8}$	**8.** $\frac{2}{9}$	**9.** $\frac{1}{2}$
10. $\frac{1}{8}$	**11.** $\frac{2}{3}$	**12.** $\frac{3}{7}$
13. $\frac{1}{9}$	**14.** $\frac{1}{4}$	**15.** $\frac{3}{5}$

Practice 15

Directions: Name each fraction. Then circle the largest fraction in each group.

Example $\dfrac{3}{7}$ $\left(\dfrac{4}{7}\right)$ $\dfrac{1}{7}$	**1.** ___ ___ ___
2. ___ ___ ___	**3.** ___ ___ ___
4. ___ ___ ___	**5.** ___ ___ ___
6. ___ ___ ___	**7.** ___ ___ ___
8. ___ ___ ___	**9.** ___ ___ ___

Practice 16

Directions: Shade the parts to show the fraction.

1.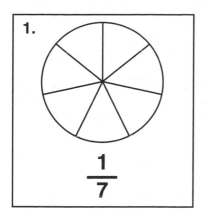
$$\frac{1}{7}$$

2.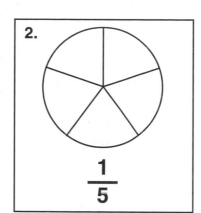
$$\frac{1}{5}$$

3.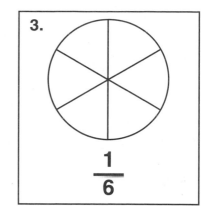
$$\frac{1}{6}$$

4.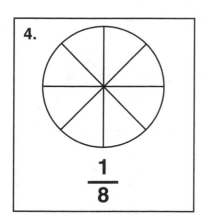
$$\frac{1}{8}$$

5.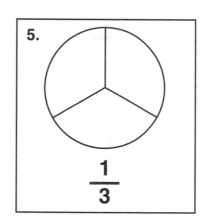
$$\frac{1}{3}$$

6.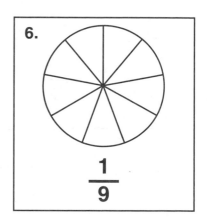
$$\frac{1}{9}$$

7.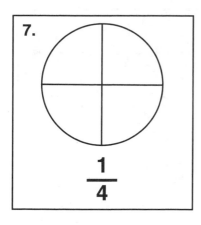
$$\frac{1}{4}$$

8.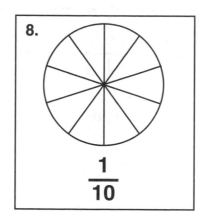
$$\frac{1}{10}$$

9.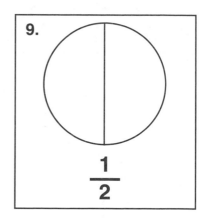
$$\frac{1}{2}$$

Directions: Write these fractions in order from largest to smallest.

_____, _____, _____, _____, _____, _____, _____, _____, _____

Practice 17

Directions: Finish each fraction sequence. Circle the fraction when the fraction amount is equal to the whole item or amount.

Example: $\frac{1}{6}$, $\frac{2}{6}$, $\frac{3}{6}$, $\frac{4}{6}$, $\frac{5}{6}$, $\boxed{\frac{6}{6}}$ ←

- The numerator has the same number as the denominator.
- When the numerator and denominator match, then the whole amount has been used.
- This is equal to 1.

1. $\frac{3}{5}$, ___, ___

2. $\frac{2}{7}$, ___, ___, ___, ___, ___

3. $\frac{1}{4}$, ___, ___, ___

4. $\frac{4}{9}$, ___, ___, ___, ___, ___

5. $\frac{5}{8}$, ___, ___, ___

6. $\frac{1}{3}$, ___, ___

7. $\frac{6}{11}$, ___, ___, ___, ___, ___

8. $\frac{4}{10}$, ___, ___, ___, ___, ___, ___

Directions: When comparing fractions with the same denominator, look at the numerator. The larger the numerator, the larger the fraction. Look at the two fractions and circle the larger fraction.

9.	$\frac{3}{4}$	$\frac{1}{4}$	10.	$\frac{2}{9}$	$\frac{4}{9}$	11.	$\frac{4}{5}$	$\frac{2}{5}$
12.	$\frac{3}{7}$	$\frac{4}{7}$	13.	$\frac{9}{10}$	$\frac{4}{10}$	14.	$\frac{7}{8}$	$\frac{6}{8}$
15.	$\frac{1}{6}$	$\frac{5}{6}$	16.	$\frac{2}{7}$	$\frac{1}{7}$	17.	$\frac{3}{10}$	$\frac{6}{10}$
18.	$\frac{6}{9}$	$\frac{1}{9}$	19.	$\frac{1}{5}$	$\frac{3}{5}$	20.	$\frac{1}{3}$	$\frac{2}{3}$
21.	$\frac{3}{8}$	$\frac{2}{8}$	22.	$\frac{2}{6}$	$\frac{3}{6}$	23.	$\frac{5}{7}$	$\frac{4}{7}$

Directions: Write the fractions in order, from smallest to largest.

24. $\frac{5}{8}$, $\frac{1}{8}$, $\frac{3}{8}$ _____, _____, _____

25. $\frac{7}{9}$, $\frac{6}{9}$, $\frac{8}{9}$ _____, _____, _____

26. $\frac{4}{7}$, $\frac{5}{7}$, $\frac{1}{7}$ _____, _____, _____

27. $\frac{3}{6}$, $\frac{5}{6}$, $\frac{2}{6}$ _____, _____, _____

28. $\frac{3}{4}$, $\frac{2}{4}$, $\frac{1}{4}$ _____, _____, _____

Practice 18

Directions: A fraction is a part of a whole item or amount. Complete each fraction sequence. Then write the fraction the arrow is pointing to on the line.

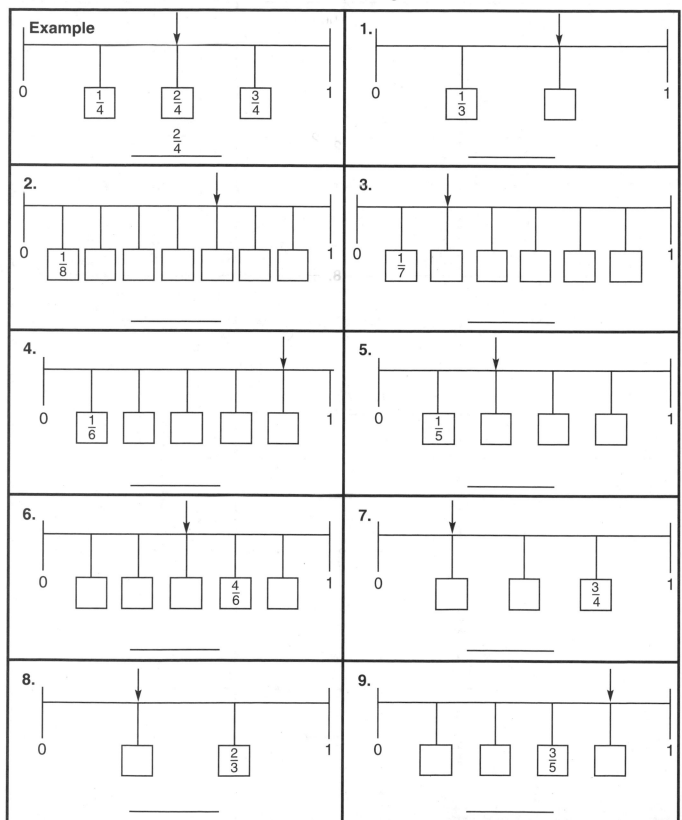

Practice 19

Directions: Identify the fraction. Circle the answer.

1.
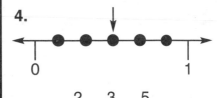

$\dfrac{1}{5}$ $\dfrac{1}{4}$ $\dfrac{1}{6}$

2.
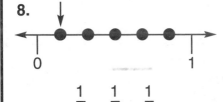

$\dfrac{2}{6}$ $\dfrac{4}{6}$ $\dfrac{5}{6}$

3.
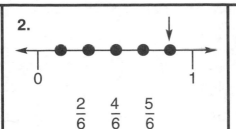

$\dfrac{1}{6}$ $\dfrac{2}{6}$ $\dfrac{3}{6}$

4.

$\dfrac{2}{6}$ $\dfrac{3}{6}$ $\dfrac{5}{6}$

5.
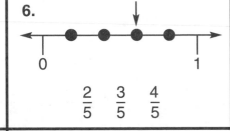

$\dfrac{1}{3}$ $\dfrac{1}{2}$ $\dfrac{3}{4}$

6.
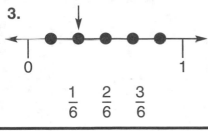

$\dfrac{2}{5}$ $\dfrac{3}{5}$ $\dfrac{4}{5}$

7.
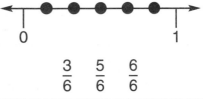

$\dfrac{2}{6}$ $\dfrac{3}{6}$ $\dfrac{4}{6}$

8.
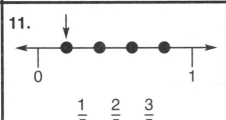

$\dfrac{1}{6}$ $\dfrac{1}{5}$ $\dfrac{1}{4}$

9.
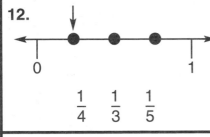

$\dfrac{1}{3}$ $\dfrac{2}{3}$

10.
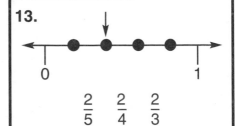

$\dfrac{3}{6}$ $\dfrac{5}{6}$ $\dfrac{6}{6}$

11.
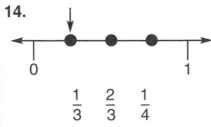

$\dfrac{1}{5}$ $\dfrac{2}{5}$ $\dfrac{3}{5}$

12.
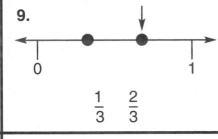

$\dfrac{1}{4}$ $\dfrac{1}{3}$ $\dfrac{1}{5}$

13.
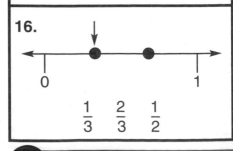

$\dfrac{2}{5}$ $\dfrac{2}{4}$ $\dfrac{2}{3}$

14.
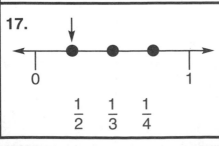

$\dfrac{1}{3}$ $\dfrac{2}{3}$ $\dfrac{1}{4}$

15.
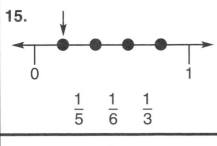

$\dfrac{1}{5}$ $\dfrac{1}{6}$ $\dfrac{1}{3}$

16.

$\dfrac{1}{3}$ $\dfrac{2}{3}$ $\dfrac{1}{2}$

17.

$\dfrac{1}{2}$ $\dfrac{1}{3}$ $\dfrac{1}{4}$

18.

$\dfrac{1}{3}$ $\dfrac{1}{4}$ $\dfrac{1}{2}$

Practice 20

Directions: Write the fraction shown by the point on the number line.

1.

2.

3.

4.

5.

6.

7.

Practice 21

Directions: Circle the correct answer.

1. $\dfrac{2}{5} + \dfrac{1}{5} =$ (A) $\dfrac{2}{25}$ (B) $\dfrac{3}{10}$ (C) $\dfrac{5}{3}$ (D) $\dfrac{3}{5}$

2. $\dfrac{3}{11} + \dfrac{1}{11} =$ (A) $\dfrac{4}{22}$ (B) $\dfrac{4}{11}$ (C) $\dfrac{11}{4}$ (D) $\dfrac{3}{121}$

3. $\dfrac{6}{11} + \dfrac{1}{11} =$ (A) $\dfrac{7}{11}$ (B) $\dfrac{6}{121}$ (C) $\dfrac{11}{7}$ (D) $\dfrac{7}{22}$

4. $\dfrac{2}{11} + \dfrac{1}{11} =$ (A) $\dfrac{11}{3}$ (B) $\dfrac{3}{11}$ (C) $\dfrac{3}{22}$ (D) $\dfrac{2}{121}$

5. $\dfrac{1}{7} + \dfrac{1}{7} =$ (A) $\dfrac{1}{49}$ (B) $\dfrac{2}{14}$ (C) $\dfrac{2}{7}$ (D) $\dfrac{7}{2}$

6. $\dfrac{1}{11} + \dfrac{1}{11} =$ (A) $\dfrac{2}{22}$ (B) $\dfrac{2}{11}$ (C) $\dfrac{1}{121}$ (D) $\dfrac{11}{2}$

7. $\dfrac{5}{7} + \dfrac{1}{7} =$ (A) $\dfrac{6}{7}$ (B) $\dfrac{5}{49}$ (C) $\dfrac{7}{6}$ (D) $\dfrac{6}{14}$

8. $\dfrac{9}{11} + \dfrac{1}{11} =$ (A) $\dfrac{11}{10}$ (B) $\dfrac{10}{11}$ (C) $\dfrac{9}{121}$ (D) $\dfrac{10}{22}$

9. $\dfrac{8}{11} + \dfrac{1}{11} =$ (A) $\dfrac{8}{121}$ (B) $\dfrac{9}{22}$ (C) $\dfrac{11}{9}$ (D) $\dfrac{9}{11}$

10. $\dfrac{1}{5} + \dfrac{1}{5} =$ (A) $\dfrac{2}{10}$ (B) $\dfrac{5}{2}$ (C) $\dfrac{2}{5}$ (D) $\dfrac{1}{25}$

Practice 22

Directions: If the denominators are the same, just add the numerators together. (The denominators stay the same.) Add the fractions. Circle the fraction sum when a whole amount is named.

1. $\frac{1}{2} + \frac{1}{2} =$ _____

2. $\frac{1}{3} + \frac{1}{3} =$ _____

3. $\frac{3}{4} + \frac{1}{4} =$ _____

4. $\frac{2}{5} + \frac{2}{5} =$ _____

5. $\frac{2}{6} + \frac{2}{6} =$ _____

6. $\frac{4}{5} + \frac{1}{5} =$ _____

7. $\frac{2}{5} + \frac{1}{5} =$ _____

8. $\frac{3}{5} + \frac{1}{5} =$ _____

9. $\frac{1}{6} + \frac{4}{6} =$ _____

10. $\frac{2}{4} + \frac{2}{4} =$ _____

11. $\frac{3}{6} + \frac{2}{6} =$ _____

12. $\frac{1}{4} + \frac{1}{4} =$ _____

13. $\frac{3}{6} + \frac{3}{6} =$ _____

14. $\frac{2}{3} + \frac{1}{3} =$ _____

15. $\frac{1}{6} + \frac{1}{6} =$ _____

16. $\frac{2}{4} + \frac{1}{4} =$ _____

17. $\frac{1}{5} + \frac{1}{5} =$ _____

18. $\frac{2}{6} + \frac{1}{6} =$ _____

19. $\frac{3}{6} + \frac{1}{6} =$ _____

20. $\frac{3}{5} + \frac{2}{5} =$ _____

21. $\frac{4}{6} + \frac{2}{6} =$ _____

Practice 23

Directions: Add the fractions.

1. $\dfrac{2}{3}$
    $+\dfrac{1}{3}$

2. $\dfrac{4}{15}$
    $+\dfrac{2}{15}$

3. $\dfrac{1}{8}$
    $+\dfrac{2}{8}$

4. $\dfrac{5}{9}$
    $+\dfrac{2}{9}$

5. $\dfrac{3}{20}$
    $+\dfrac{4}{20}$

6. $\dfrac{2}{7}$
    $+\dfrac{1}{7}$

7. $\dfrac{3}{10}$
    $+\dfrac{5}{10}$

8. $\dfrac{4}{5}$
    $+\dfrac{1}{5}$

9. $\dfrac{7}{24}$
    $+\dfrac{10}{24}$

10. $\dfrac{4}{7}$
    $+\dfrac{1}{7}$

11. $\dfrac{3}{5}$
    $+\dfrac{1}{5}$

12. $\dfrac{5}{8}$
    $+\dfrac{2}{8}$

13. $\dfrac{6}{10}$
    $+\dfrac{1}{10}$

14. $\dfrac{1}{3}$
    $+\dfrac{1}{3}$

15. $\dfrac{10}{13}$
    $+\dfrac{1}{13}$

16. $\dfrac{5}{12}$
    $+\dfrac{3}{12}$

Practice 24 ⟳ ⟳ ⟳ ⟳ ⟳ ⟳ ⟳ ⟳ ⟳ ⟳ ⟳ ⟳ ⟳ ⟳

Directions: Circle the correct answer.

1. $\dfrac{5}{7} - \dfrac{1}{7} =$ (A) 4   (B) $\dfrac{4}{7}$ (C) $\dfrac{6}{7}$ (D) $\dfrac{5}{7}$

2. $\dfrac{3}{17} - \dfrac{2}{17} =$ (A) 1   (B) $\dfrac{2}{17}$ (C) $\dfrac{1}{17}$ (D) $\dfrac{5}{17}$

3. $\dfrac{12}{13} - \dfrac{11}{13} =$ (A) $\dfrac{1}{13}$ (B) $\dfrac{2}{13}$ (C) $\dfrac{23}{13}$ (D) 1

4. $\dfrac{5}{13} - \dfrac{2}{13} =$ (A) 3   (B) $\dfrac{4}{13}$ (C) $\dfrac{7}{13}$ (D) $\dfrac{3}{13}$

5. $\dfrac{8}{13} - \dfrac{7}{13} =$ (A) $\dfrac{2}{13}$ (B) $\dfrac{15}{13}$ (C) $\dfrac{1}{13}$ (D) 1

6. $\dfrac{6}{11} - \dfrac{1}{11} =$ (A) $\dfrac{7}{11}$ (B) 5   (C) $\dfrac{6}{11}$ (D) $\dfrac{5}{11}$

7. At lunch, Harriet had $\dfrac{5}{8}$ cup of popcorn. She then gave $\dfrac{1}{8}$ cup of popcorn to her friend. How many cups of popcorn does she have left?

    (A) $\dfrac{4}{8}$ cup   (B) $\dfrac{5}{8}$ cup   (C) $\dfrac{16}{6}$ cups   (D) $\dfrac{4}{16}$ cup

8. Juanita has $\dfrac{3}{8}$ cup of raisins. She needs $\dfrac{1}{8}$ cup of raisins for her cookie recipe. How many cups of raisins will be left after she makes her cookies?

    (A) $\dfrac{2}{16}$ cup   (B) $\dfrac{3}{8}$ cup   (C) $\dfrac{2}{8}$ cup   (D) $\dfrac{16}{4}$ cups

Practice 25

Directions: If the denominators are the same, just subtract the second numerator from the first numerator. (The denominators stay the same.) Subtract the fractions.

1. $\frac{4}{6} - \frac{2}{6} =$ _____

2. $\frac{1}{5} - \frac{1}{5} =$ _____

3. $\frac{2}{3} - \frac{1}{3}$ _____

4. $\frac{3}{5} - \frac{1}{5} =$ _____

5. $\frac{2}{6} - \frac{1}{6}$ _____

6. $\frac{2}{4} - \frac{1}{4} =$ _____

7. $\frac{3}{4} - \frac{1}{4} =$ _____

8. $\frac{3}{6} - \frac{2}{6}$ _____

9. $\frac{4}{6} - \frac{1}{6} =$ _____

10. $\frac{4}{5} - \frac{1}{5} =$ _____

11. $\frac{3}{6} - \frac{3}{6} =$ _____

12. $\frac{2}{5} - \frac{1}{5} =$ _____

13. $\frac{3}{6} - \frac{1}{6} =$ _____

14. $\frac{7}{8} - \frac{4}{8} =$ _____

15. $\frac{5}{7} - \frac{1}{7} =$ _____

16. $\frac{6}{8} - \frac{1}{8} =$ _____

17. $\frac{4}{7} - \frac{2}{7}$ _____

18. $\frac{7}{9} - \frac{6}{9} =$ _____

19. $\frac{8}{9} - \frac{1}{9} =$ _____

20. $\frac{4}{9} - \frac{2}{9} =$ _____

21. $\frac{6}{7} - \frac{5}{7} =$ _____

Practice 28

Directions: Shade to show the correct fraction. Then circle the larger fraction.

Example	1.
$\frac{9}{12}$ $\frac{1}{2}$	$\frac{3}{9}$ $\frac{4}{10}$

2.	3.
$\frac{8}{10}$ $\frac{4}{6}$	$\frac{1}{4}$ $\frac{2}{3}$

4.	5.
$\frac{8}{9}$ $\frac{7}{8}$	$\frac{2}{7}$ $\frac{5}{9}$

6.	7.
$\frac{4}{7}$ $\frac{3}{10}$	$\frac{3}{7}$ $\frac{1}{6}$

8.	9.
$\frac{3}{4}$ $\frac{1}{5}$	$\frac{3}{8}$ $\frac{1}{9}$

Practice 31

---

1. Write the fraction for the shaded part.

_____

2. Write the fraction for the shaded part.

_____

3. Write the fraction for the shaded part.

_____

---

4. Write the fraction. Write < or > to compare the fractions.

_____ ◯ _____

5. Write the fraction. Write < or > to compare the fractions.

_____ ◯ _____

---

6. Write the fraction for the shaded part.

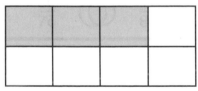

_____

7. Write the fraction for the shaded part.

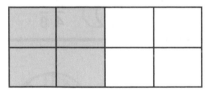

_____

8. Write the fraction for the shaded part.

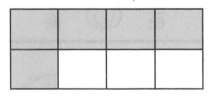

_____

---

9. Write the fraction. Write < or > to compare the fractions.

_____ ◯ _____

10. Write the fraction. Write < or > to compare the fractions.

 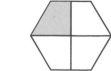

_____ ◯ _____

Practice 32

Directions: How many equal sets can be made? Circle the equal sets. Write the fraction for one set out of all of the sets.

Example  Circle sets of 2. ___7___ equal sets can be made. The fraction is ___$\frac{1}{7}$___ .	**1.**  Circle sets of 4. _____ equal sets can be made. The fraction is _____ .
2.  Circle sets of 6. _____ equal sets can be made. The fraction is _____ .	**3.**  Circle sets of 3. _____ equal sets can be made. The fraction is _____ .
4.  Circle sets of 7. _____ equal sets can be made. The fraction is _____ .	**5.**  Circle sets of 5. _____ equal sets can be made. The fraction is _____ .
6.  Circle sets of 1. _____ equal sets can be made. The fraction is _____ .	**7.** Circle sets of 5. _____ equal sets can be made. The fraction is _____ .

Practice 33

Directions: How many equal sets can be made? Circle equal sets. Write the fraction for one set out of all of the sets.

1.

Circle sets of 3.

_____ equal sets can be made.

The fraction is _____ .

2.

Circle sets of 2.

_____ equal sets can be made.

The fraction is _____ .

3.

Circle sets of 2.

_____ equal sets can be made.

The fraction is _____ .

4.

Circle sets of 1.

_____ equal sets can be made.

The fraction is _____ .

5.

Circle sets of 1.

_____ equal sets can be made.

The fraction is _____ .

6.

Circle sets of 4.

_____ equal sets can be made.

The fraction is _____ .

7.

Circle sets of 3.

_____ equal sets can be made.

The fraction is _____ .

8.

Circle sets of 4.

_____ equal sets can be made.

The fraction is _____ .

Practice 34

The top is the **numerator**. It tells how many items or parts of the whole set is used or needed.

$\frac{2}{5}$

The bottom number is the **denominator**. It tells how many equal parts are in the set.

There are 5 equal parts in the set. Each set has 1 bird. $\frac{2}{5}$ of the set is 2 birds.

Directions: Circle the fractional amount. Complete the sentence.

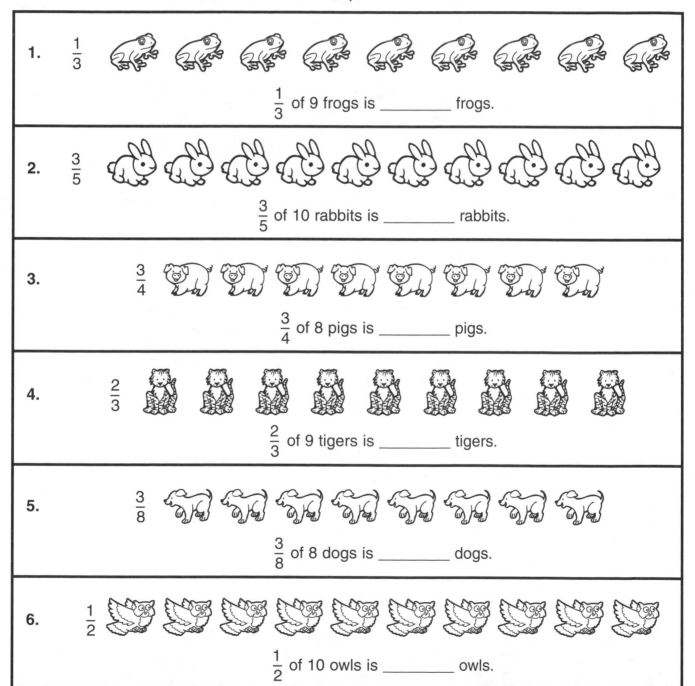

1. $\frac{1}{3}$

$\frac{1}{3}$ of 9 frogs is _____ frogs.

2. $\frac{3}{5}$

$\frac{3}{5}$ of 10 rabbits is _____ rabbits.

3. $\frac{3}{4}$

$\frac{3}{4}$ of 8 pigs is _____ pigs.

4. $\frac{2}{3}$

$\frac{2}{3}$ of 9 tigers is _____ tigers.

5. $\frac{3}{8}$

$\frac{3}{8}$ of 8 dogs is _____ dogs.

6. $\frac{1}{2}$

$\frac{1}{2}$ of 10 owls is _____ owls.

Practice 35

Money can be written as a fraction.

1¢	is the same as	$\dfrac{1}{100}$	$0.41	is the same as	$\dfrac{41}{100}$
It takes 100 cents to make a dollar. One penny is one-hundredth of a dollar.			It takes 100 cents to make a dollar. 41¢ is 41 hundredths of a dollar.		

Directions: Write each amount of money as a fraction. The first one has already been done for you.

1.

$$9¢ = \frac{9}{100}$$

2.

$$44¢ = \underline{\quad}$$

3.

$$98¢ = \underline{\quad}$$

4.

$$65¢ = \underline{\quad}$$

5.

$$39¢ = \underline{\quad}$$

6.

$$27¢ = \underline{\quad}$$

7.

$$\$0.73 = \underline{\quad}$$

8.

$$\$0.88 = \underline{\quad}$$

9.

$$\$0.23 = \underline{\quad}$$

10.

$$\$0.15 = \underline{\quad}$$

11.

$$\$0.50 = \underline{\quad}$$

12.

$$\$0.11 = \underline{\quad}$$

Directions: Use the > (greater than), < (less than), or = (equal to) symbols to compare the numbers. The first one has already been done for you.

13.

$$46¢ \; \boxed{<} \; \frac{99}{100}$$

14.

$$57¢ \; \bigcirc \; \frac{83}{100}$$

15.

$$18¢ \; \bigcirc \; \frac{18}{100}$$

16.

$$\$0.25 \; \bigcirc \; \frac{20}{100}$$

17.

$$\$0.63 \; \bigcirc \; \frac{63}{100}$$

18.

$$\$0.74 \; \bigcirc \; \frac{31}{100}$$

Practice 36

Dollars and cents can be written as fractions, too.

$1.00 is the same as $\frac{100}{100}$ or $\frac{1}{1}$ or 1	$1.23 is the same as $\frac{123}{100}$ or $1\frac{23}{100}$

Directions: Write each amount of money as a mixed number. The first one has already been done for you.

1. $5.81 = 5\frac{81}{100}$

2. $2.71 = ———

3. $3.19 = ———

4. $1.86 = ———

5. $4.24 = ———

6. $5.11 = ———

7. $1.63 = ———

8. $2.10 = ———

9. $3.17 = ———

10. $4.29 = ———

11. $7.99 = ———

12. $3.00 = ———

Directions: Convert each fraction to money. The first one has already been done for you.

13. $6\frac{19}{100}$ = $6.19

14. $2\frac{34}{100}$ =

15. $4\frac{98}{100}$ =

16. $3\frac{68}{100}$ =

17. $1\frac{50}{100}$ =

18. $2\frac{27}{100}$ =

Practice 37

In a **circle graph**, all the parts must add up to be a whole. Think of the parts like pieces that add up to one whole pie. Look at these pies and how they are divided into pieces.

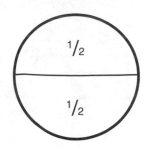

1/2 a pie

+ 1/2 a pie

2 halves =

1 whole pie

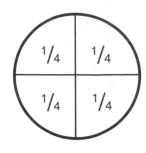

1/4 a pie

+ 1/4 a pie

+ 1/4 a pie

+ 1/4 a pie

4 fourths =

1 whole pie

Directions: Make a circle graph to show how much pie a family ate. Here is the information you will need.

Mother ate 1/4 of the pie.

Sister ate 1/4 of the pie.

Father ate 1/4 of the pie.

Brother ate 1/8 of the pie.

Grandma ate 1/8 of the pie.

Directions: Color the graph below using the Color Key.

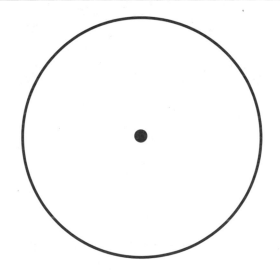

Pie My Family Ate

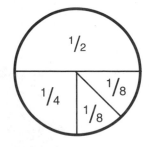

1/8 a pie

+ 1/8 a pie

+ 1/8 a pie

+ 1/8 a pie

+ 1/8 a pie

+ 1/8 a pie

+ 1/8 a pie

+ 1/8 a pie

8 eighths =

1 whole pie

1/2 a pie = 1 half

+ 1/4 a pie = 1 fourth

+ 1/8 a pie = 1 eighth

+ 1/8 a pie = 1 eighth

1 whole pie

Color Key

Sister = orange   Mother = pink

Grandma = red   Brother = yellow

Father = blue

Test Practice 3

Directions: Find the correct answer.

1. What's the fraction? (A) $\frac{1}{4}$ (B) $\frac{3}{4}$ (C) $\frac{2}{4}$	**2.** What's the fraction? (A) $\frac{3}{5}$ (B) $\frac{2}{5}$ (C) $\frac{2}{3}$	**3.** What's the fraction? 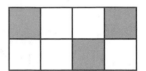 (A) $\frac{7}{8}$ (B) $\frac{3}{8}$ (C) $\frac{1}{8}$
4. Name the largest fraction. (A) $\frac{3}{7}$ (B) $\frac{2}{7}$ (C) $\frac{4}{7}$	**5.** Name the largest fraction. (A) $\frac{2}{6}$ (B) $\frac{1}{6}$ (C) $\frac{4}{6}$	**6.** Name the smallest fraction. (A) $\frac{5}{9}$ (B) $\frac{3}{9}$ (C) $\frac{1}{9}$
7. Which fraction names a whole item or amount? (A) $\frac{1}{8}$ (B) $\frac{8}{8}$ (C) $\frac{3}{8}$	**8.** Add the fractions. $\frac{1}{4}+\frac{2}{4}$ (A) $\frac{3}{4}$ (B) $\frac{2}{16}$ (C) $\frac{3}{8}$	**9.** Add the fractions. $\frac{3}{5}+\frac{1}{5}$ (A) $\frac{4}{10}$ (B) $\frac{4}{5}$ (C) $\frac{4}{25}$
10. Subtract the fractions. $\frac{5}{8}-\frac{3}{8}$ (A) $\frac{2}{0}$ (B) $\frac{8}{8}$ (C) $\frac{2}{8}$	**11.** Subtract the fractions. $\frac{7}{9}-\frac{6}{9}$ (A) $\frac{9}{1}$ (B) $\frac{13}{9}$ (C) $\frac{1}{9}$	**12.** Subtract the fractions. $\frac{6}{7}-\frac{4}{7}$ (A) $\frac{2}{7}$ (B) $\frac{10}{7}$ (C) $\frac{2}{0}$

13. Name the fraction that the arrows points to.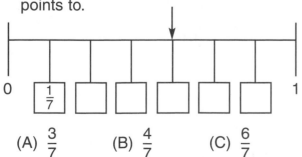
(A) $\frac{3}{7}$ (B) $\frac{4}{7}$ (C) $\frac{6}{7}$

14. Name the fraction that the arrows points to.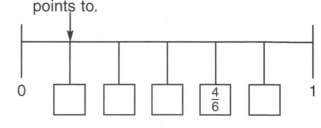
(A) $\frac{1}{6}$ (B) $\frac{2}{6}$ (C) $\frac{3}{6}$

Test Practice 4

Directions: Find the correct answer to each problem. Choose *none of these* if the right answer is not given.

1.

$$\frac{9}{11} - \frac{2}{11} =$$

(A) $\frac{2}{11}$ (C) $\frac{7}{11}$

(B) $\frac{5}{11}$ (D) $\frac{9}{11}$

(E) none of these

2.

$$\frac{3}{5} + \frac{2}{5} =$$

(A) $\frac{1}{10}$ (C) $\frac{5}{5}$

(B) $\frac{1}{5}$ (D) $\frac{5}{10}$

(E) none of these

3.

$$\frac{5}{8} + \frac{1}{8} =$$

(A) $\frac{4}{8}$ (C) $\frac{6}{8}$

(B) $\frac{1}{16}$ (D) $\frac{6}{16}$

(E) none of these

4.

$$\frac{2}{3} - \frac{1}{3} =$$

(A) $\frac{1}{3}$ (C) 1

(B) $\frac{1}{6}$ (D) 3

(E) none of these

5.

$$\frac{7}{12} - \frac{6}{12} =$$

(A) $\frac{1}{24}$ (C) $\frac{13}{12}$

(B) $\frac{1}{12}$ (D) $\frac{13}{24}$

(E) none of these

6.

$$\begin{array}{r} \frac{11}{13} \\ - \ \frac{5}{13} \\ \hline \end{array}$$

(A) $\frac{7}{13}$ (C) $\frac{4}{13}$

(B) $\frac{16}{13}$ (D) $\frac{6}{13}$

(E) none of these

7.

$$\begin{array}{r} \frac{6}{7} \\ - \ \frac{2}{7} \\ \hline \end{array}$$

(A) $\frac{4}{7}$ (C) $\frac{4}{14}$

(B) $\frac{8}{7}$ (D) $\frac{12}{14}$

(E) none of these

8.

$$\begin{array}{r} \frac{7}{8} \\ - \ \frac{4}{8} \\ \hline \end{array}$$

(A) $\frac{1}{8}$ (C) $\frac{11}{8}$

(B) $\frac{3}{8}$ (D) 8

(E) none of these

9.

$$\begin{array}{r} \frac{2}{10} \\ + \ \frac{1}{10} \\ \hline \end{array}$$

(A) $\frac{1}{10}$ (C) $\frac{3}{10}$

(B) $\frac{3}{7}$ (D) $\frac{4}{10}$

(E) none of these

10.

$$\begin{array}{r} \frac{4}{8} \\ + \ \frac{3}{8} \\ \hline \end{array}$$

(A) $\frac{7}{8}$ (C) $\frac{1}{2}$

(B) $\frac{2}{8}$ (D) $\frac{1}{8}$

(E) none of these

Test Practice 5

Directions: Find the correct answer.

---

1. Find the fraction using words.

(A) one-half   (B) one-third   (C) one-fourth

2. Find the fraction using words.

(A) one-sixth   (B) one-seventh   (C) one-eighth

---

3. Compare the fractions.

$\frac{1}{2}$ ◯ $\frac{3}{4}$

(A) >   (B) <   (C) =

4. Compare the fractions.

$\frac{4}{8}$ ◯ $\frac{3}{6}$

(A) >   (B) <   (C) =

---

5. Divide the pictures into 2 equal sets. Find the correct answer.

$\frac{1}{2}$ of 10 ladybugs is _____ ladybugs.

(A) 4   (B) 2   (C) 3   (D) 5

6. Divide the pictures into 3 equal sets. Find the correct answer.

$\frac{1}{3}$ of 9 bees is _____ bees.

(A) 3   (B) 4   (C) 2   (D) 5

---

7. Find the fraction.

53¢ = _____

(A) $3\frac{5}{10}$ (B) $5\frac{3}{10}$ (C) $\frac{53}{10}$ (D) $\frac{53}{100}$

8. Find the mixed number.

$3.61 = _____

(A) $36\frac{1}{100}$ (B) $3\frac{61}{100}$ (C) $3\frac{1}{61}$ (D) $\frac{3}{61}$

Answer Sheet

Test Practice 1

1. (A) (B) (C)
2. (A) (B) (C)
3. (A) (B)
4. (A) (B) (C)
5. (A) (B) (C)
6. (A) (B) (C)
7. (A) (B) (C)
8. (A) (B) (C)
9. (A) (B) (C)
10. (A) (B) (C)

Test Practice 2

1. (A) (B) (C) (D)
2. (A) (B) (C) (D)
3. (A) (B) (C) (D)
4. (A) (B) (C) (D)
5. (A) (B) (C) (D)
6. (A) (B) (C) (D)

Test Practice 3

1. (A) (B) (C)
2. (A) (B) (C)
3. (A) (B) (C)
4. (A) (B) (C)
5. (A) (B) (C)
6. (A) (B) (C)
7. (A) (B) (C)
8. (A) (B) (C)
9. (A) (B) (C)
10. (A) (B) (C)
11. (A) (B) (C)
12. (A) (B) (C)
13. (A) (B) (C)
14. (A) (B) (C)

Test Practice 4

1. (A) (B) (C) (D) (E)
2. (A) (B) (C) (D) (E)
3. (A) (B) (C) (D) (E)
4. (A) (B) (C) (D) (E)
5. (A) (B) (C) (D) (E)
6. (A) (B) (C) (D) (E)
7. (A) (B) (C) (D) (E)
8. (A) (B) (C) (D) (E)
9. (A) (B) (C) (D) (E)
10. (A) (B) (C) (D) (E)

Test Practice 5

1. (A) (B) (C)
2. (A) (B) (C)
3. (A) (B) (C)
4. (A) (B) (C)
5. (A) (B) (C) (D)
6. (A) (B) (C) (D)
7. (A) (B) (C) (D)
8. (A) (B) (C) (D)

#8601 Practice Makes Perfect: Fractions

Answer Key

Page 4
1. 2	7. 5
2. 5	8. 8
3. 4	9. 4
4. 3	10. 8
5. 6	11. 7
6. 2	12. 10

Page 5
1. no	7. yes
2. no	8. no
3. yes	9. no
4. no	10. no
5. no	11. yes
6. yes	12. no

Page 6
1. circle with four parts, triangle with four parts
2. circle with three parts
3. heart with two parts, triangle with two parts
4. rectangle with five parts, star with five parts
5. rectangle with six parts

Page 7
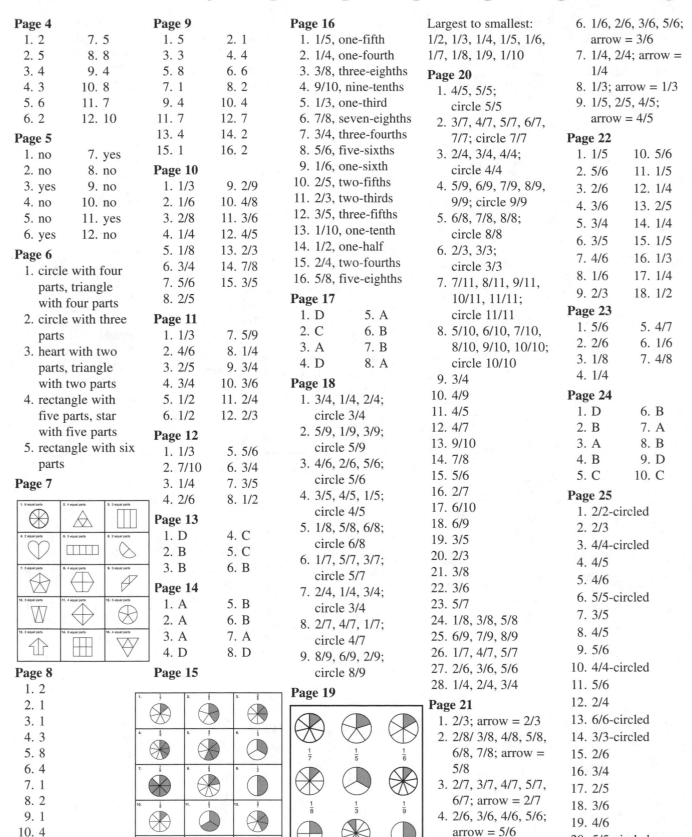

Page 8
1. 2
2. 1
3. 1
4. 3
5. 8
6. 4
7. 1
8. 2
9. 1
10. 4
11. 1
12. 2

Page 9
1. 5	2. 1
3. 3	4. 4
5. 8	6. 6
7. 1	8. 2
9. 4	10. 4
11. 7	12. 7
13. 4	14. 2
15. 1	16. 2

Page 10
1. 1/3	9. 2/9
2. 1/6	10. 4/8
3. 2/8	11. 3/6
4. 1/4	12. 4/5
5. 1/8	13. 2/3
6. 3/4	14. 7/8
7. 5/6	15. 3/5
8. 2/5	

Page 11
1. 1/3	7. 5/9
2. 4/6	8. 1/4
3. 2/5	9. 3/4
4. 3/4	10. 3/6
5. 1/2	11. 2/4
6. 1/2	12. 2/3

Page 12
1. 1/3	5. 5/6
2. 7/10	6. 3/4
3. 1/4	7. 3/5
4. 2/6	8. 1/2

Page 13
1. D	4. C
2. B	5. C
3. B	6. B

Page 14
1. A	5. B
2. A	6. B
3. A	7. A
4. D	8. D

Page 15
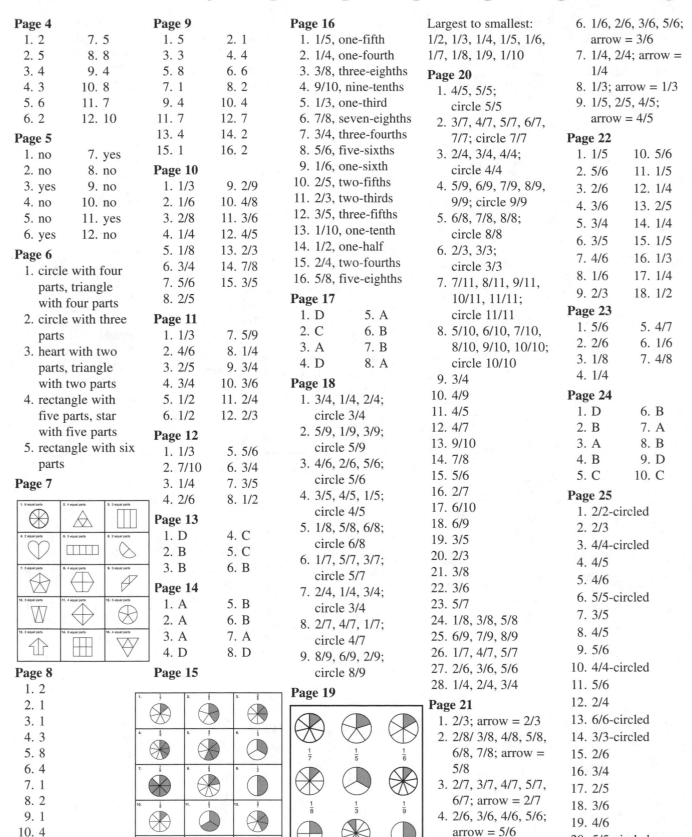

Page 16
1. 1/5, one-fifth
2. 1/4, one-fourth
3. 3/8, three-eighths
4. 9/10, nine-tenths
5. 1/3, one-third
6. 7/8, seven-eighths
7. 3/4, three-fourths
8. 5/6, five-sixths
9. 1/6, one-sixth
10. 2/5, two-fifths
11. 2/3, two-thirds
12. 3/5, three-fifths
13. 1/10, one-tenth
14. 1/2, one-half
15. 2/4, two-fourths
16. 5/8, five-eighths

Page 17
1. D	5. A
2. C	6. B
3. A	7. B
4. D	8. A

Page 18
1. 3/4, 1/4, 2/4; circle 3/4
2. 5/9, 1/9, 3/9; circle 5/9
3. 4/6, 2/6, 5/6; circle 5/6
4. 3/5, 4/5, 1/5; circle 4/5
5. 1/8, 5/8, 6/8; circle 6/8
6. 1/7, 5/7, 3/7; circle 5/7
7. 2/4, 1/4, 3/4; circle 3/4
8. 2/7, 4/7, 1/7; circle 4/7
9. 8/9, 6/9, 2/9; circle 8/9

Page 19
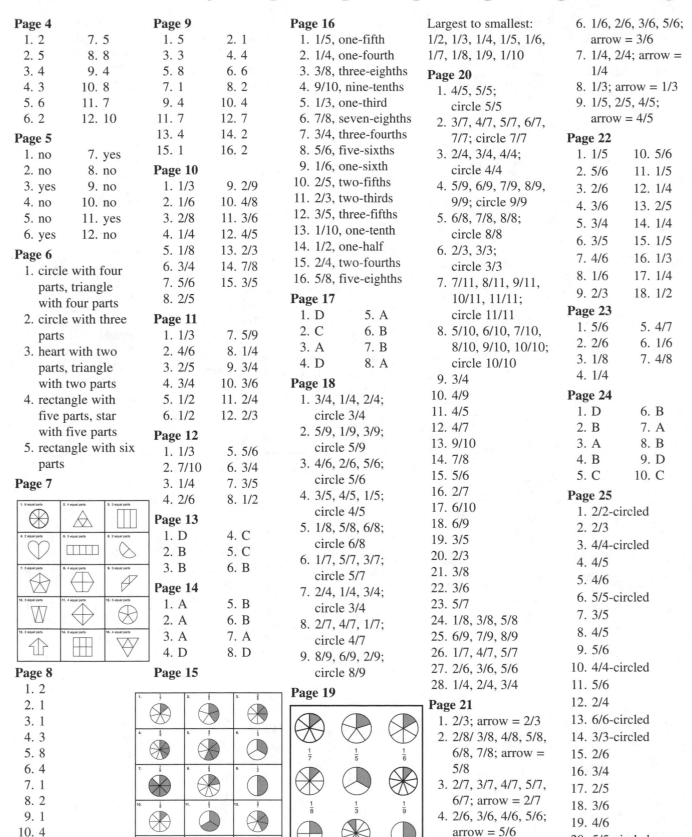

Page 20
1. 4/5, 5/5; circle 5/5
2. 3/7, 4/7, 5/7, 6/7, 7/7; circle 7/7
3. 2/4, 3/4, 4/4; circle 4/4
4. 5/9, 6/9, 7/9, 8/9, 9/9; circle 9/9
5. 6/8, 7/8, 8/8; circle 8/8
6. 2/3, 3/3; circle 3/3
7. 7/11, 8/11, 9/11, 10/11, 11/11; circle 11/11
8. 5/10, 6/10, 7/10, 8/10, 9/10, 10/10; circle 10/10
9. 3/4
10. 4/9
11. 4/5
12. 4/7
13. 9/10
14. 7/8
15. 5/6
16. 2/7
17. 6/10
18. 6/9
19. 3/5
20. 2/3
21. 3/8
22. 3/6
23. 5/7
24. 1/8, 3/8, 5/8
25. 6/9, 7/9, 8/9
26. 1/7, 4/7, 5/7
27. 2/6, 3/6, 5/6
28. 1/4, 2/4, 3/4

Page 21
1. 2/3; arrow = 2/3
2. 2/8/ 3/8, 4/8, 5/8, 6/8, 7/8; arrow = 5/8
3. 2/7, 3/7, 4/7, 5/7, 6/7; arrow = 2/7
4. 2/6, 3/6, 4/6, 5/6; arrow = 5/6
5. 2/5, 3/5, 4/5; arrow = 2/5

Largest to smallest:
1/2, 1/3, 1/4, 1/5, 1/6, 1/7, 1/8, 1/9, 1/10

6. 1/6, 2/6, 3/6, 5/6; arrow = 3/6
7. 1/4, 2/4; arrow = 1/4
8. 1/3; arrow = 1/3
9. 1/5, 2/5, 4/5; arrow = 4/5

Page 22
1. 1/5	10. 5/6
2. 5/6	11. 1/5
3. 2/6	12. 1/4
4. 3/6	13. 2/5
5. 3/4	14. 1/4
6. 3/5	15. 1/5
7. 4/6	16. 1/3
8. 1/6	17. 1/4
9. 2/3	18. 1/2

Page 23
1. 5/6	5. 4/7
2. 2/6	6. 1/6
3. 1/8	7. 4/8
4. 1/4	

Page 24
1. D	6. B
2. B	7. A
3. A	8. B
4. B	9. D
5. C	10. C

Page 25
1. 2/2-circled
2. 2/3
3. 4/4-circled
4. 4/5
5. 4/6
6. 5/5-circled
7. 3/5
8. 4/5
9. 5/6
10. 4/4-circled
11. 5/6
12. 2/4
13. 6/6-circled
14. 3/3-circled
15. 2/6
16. 3/4
17. 2/5
18. 3/6
19. 4/6
20. 5/5-circled
21. 6/6-circled

Answer Key

Page 26
1. 3/3
2. 6/15
3. 3/8
4. 7/9
5. 7/20
6. 3/7
7. 8/10
8. 5/5
9. 17/24
10. 5/7
11. 4/5
12. 7/8
13. 7/10
14. 2/3
15. 11/13
16. 8/12

Page 27
1. B
2. C
3. A
4. D
5. C
6. D
7. A
8. C

Page 28
1. 2/6
2. 0/5
3. 1/3
4. 2/5
5. 1/6
6. 1/4
7. 2/4
8. 1/6
9. 3/6
10. 3/5
11. 0/6
12. 1/5
13. 2/6
14. 3/8
15. 4/7
16. 5/8
17. 2/7
18. 1/9
19. 7/9
20. 2/9
21. 1/7

Page 29
1. 1/12
2. 5/8
3. 6/16
4. 2/6
5. 4/6
6. 0/3
7. 8/10
8. 3/8
9. 2/4
10. 1/5
11. 8/11
12. 6/10
13. 1/10
14. 4/16
15. 4/12
16. 1/14

Page 30
1. 1/4
2. 3/5
3. 5/6
4. 8/9
5. 2/7
6. 4/9
7. 3/4
8. 5/9
9. 1/3
10. 4/8
11. 2/9
12. 2/8
13. 5/6
14. 4/7
15. 7/9
16. 2/4
17. 3/5
18. 2/3
19. 4/5

Page 31

Page 32

Page 33

Page 34
1. 1/2
2. 2/3
3. 2/4
4. 1/2 < 3/4
5. 1/4 < 1/2
6. 3/8
7. 4/8
8. 5/8
9. 2/5 < 3/5
10. 2/4 > 1/4

Page 35
1. Make sure sets of
   4 are circled.
   3 equal sets can
   be made.
   The fraction is 1/3
2. Make sure sets of
   6 are circled.

2 equal sets can
be made.
The fraction is 1/2.
3. Make sure sets of
   3 are circled.
   3 equal sets can
   be made.
   The fraction is 1/3.
4. Make sure sets of
   7 are circled.
   2 equal sets can
   be made.
   The fraction is 1/2.
5. Make sure sets of
   5 are circled.
   2 equal sets can
   be made.
   The fraction is 1/2.
6. Make sure sets of
   1 are circled.
   12 equal sets can
   be made.
   The fraction is 1/12.
7. Make sure sets of
   5 are circled.
   3 equal sets can
   be made.
   The fraction is 1/3.

Page 36
1. Make sure sets of
   3 are circled.
   4 equal sets can
   be made.
   The fraction is 1/4
2. Make sure sets of
   2 are circled.
   6 equal sets can
   be made.
   The fraction is 1/6.
3. Make sure sets of
   2 are circled.
   5 equal sets can
   be made.
   The fraction is 1/5.
4. Make sure sets of
   1 are circled.
   8 equal sets can
   be made.
   The fraction is 1/8.
5. Make sure sets of
   1 are circled.
   10 equal sets can
   be made.

The fraction is 1/10.
6. Make sure sets of
   4 are circled.
   4 equal sets can
   be made.
   The fraction is 1/4.
7. Make sure sets of
   3 are circled.
   3 equal sets can
   be made.
   The fraction is 1/3.
8. Make sure sets of
   4 are circled.
   3 equal sets can
   be made.
   The fraction is 1/3.

Page 37
1. 3 (3 frogs circled)
2. 6 (6 rabbits
   circled)
3. 6 (6 pigs circled)
4. 6 (6 tigers
   circled)
5. 3 (3 dogs circled)
6. 5 (5 owls circled)

Page 38
1. 9/100
2. 44/100
3. 98/100
4. 65/100
5. 39/100
6. 27/100
7. 73/100
8. 88/100
9. 23/100
10. 15/100
11. 50/100
12. 11/100
13. <
14. <
15. =
16. >
17. =
18. >

Page 39
1. 5 81/100
2. 2 71/100
3. 3 19/100
4. 1 86/100
5. 4 24/100
6. 5 11/100
7. 1 63/100

8. 2 10/100
9. 3 17/100
10. 4 29/100
11. 7 99/100
12. 300/100 or 3
13. $6.19
14. $2.34
15. $4.98
16. $3.68
17. $1.50
18. $2.27

Page 40

Page 41
1. B
2. C
3. A
4. B
5. A
6. A
7. B
8. A
9. B
10. A

Page 42
1. B
2. C
3. B
4. B
5. D
6. A

Page 43
1. B
2. A
3. B
4. C
5. C
6. C
7. B
8. A
9. B
10. C
11. C
12. A
13. B
14. A

Page 44
1. C
2. C
3. C
4. A
5. B
6. D
7. A
8. B
9. C
10. A

Page 45
1. B
2. C
3. B
4. C
5. D
6. A
7. D
8. B

 #8601 Practice Makes Perfect: Fractions